I0483734

NIST Special Publication 800-127

Guide to Securing WiMAX Wireless Communications

Recommendations of the National Institute of Standards and Technology

Karen Scarfone
Cyrus Tibbs
Matthew Sexton

COMPUTER SECURITY

Computer Security Division
Information Technology Laboratory
National Institute of Standards and Technology
Gaithersburg, MD 20899-8930

September 2010

U.S. Department of Commerce

Gary Locke, Secretary

National Institute of Standards and Technology

Patrick D. Gallagher, Director

Reports on Computer Systems Technology

The Information Technology Laboratory (ITL) at the National Institute of Standards and Technology (NIST) promotes the U.S. economy and public welfare by providing technical leadership for the nation's measurement and standards infrastructure. ITL develops tests, test methods, reference data, proof of concept implementations, and technical analysis to advance the development and productive use of information technology. ITL's responsibilities include the development of technical, physical, administrative, and management standards and guidelines for the cost-effective security and privacy of sensitive unclassified information in Federal computer systems. This Special Publication 800-series reports on ITL's research, guidance, and outreach efforts in computer security and its collaborative activities with industry, government, and academic organizations.

National Institute of Standards and Technology Special Publication 800-127
Natl. Inst. Stand. Technol. Spec. Publ. 800-127, 44 pages (September 2010)

Acknowledgments

The authors, Karen Scarfone of the National Institute of Standards and Technology (NIST) and Cyrus Tibbs and Matthew Sexton of Booz Allen Hamilton, wish to thank their colleagues who reviewed drafts of this document and contributed to its technical content, particularly Lily Chen and Tim Grance of NIST and Bernard Eydt, Ross Schwalm, and John Padgette of Booz Allen Hamilton. The authors would also like to thank the WiMAX Forum for their contributions to this publication.

Trademark Information

WiMAX, WiMAX Forum, WiMAX Certified, and WiMAX Forum Certified are trademarks or registered trademarks of the WiMAX Forum.

All other names are trademarks or registered trademarks of their respective owners.

Table of Contents

List of Appendices

List of Figures

List of Tables

Executive Summary

WiMAX[1] technology is a wireless metropolitan area network (WMAN) communications technology that is largely based on the wireless interface defined in the IEEE 802.16 standard. The industry trade association, the WiMAX Forum, coined the WiMAX trademark and defines the precise content and scope of WiMAX technology through technical specifications that it creates and publishes.[2] The original purpose of IEEE 802.16 technology was to provide last-mile broadband wireless access as an alternative to cable, digital subscriber line-, or T1 service. Developments in the IEEE 802.16 standard shifted the technology's focus toward a more cellular-like, mobile architecture to serve a broader market. Today, WiMAX technology continues to adapt to market demands and provide enhanced user mobility. This document discusses WiMAX wireless communication topologies, components, certifications, security features, and related security concerns.

The IEEE amendment that enabled mobile WiMAX operations is IEEE 802.16e-2005. Prior to its release, deployment of WiMAX networks was limited to fixed operations by the IEEE 802.16-2004 standard. Additionally, IEEE 802.16e-2005 provided significant security enhancements to its predecessor by incorporating more robust mutual authentication mechanisms, as well as support for Advanced Encryption Standard (AES). Although the IEEE 802.16-2004 and 802.16e-2005 standards were released within a year of each other, IEEE 802.16e-2005 product certification did not start until 2008, and IEEE 802.16-2004 products are still used in today's information technology (IT) environments. The most recently ratified standard is IEEE 802.16-2009, which consolidated IEEE 802.16-2004, IEEE 802.16e-2005, and other IEEE 802.16 amendments from 2004 through 2008. IEEE also released IEEE 802.16j-2009 to specify multi-hop relay networking. This publication addresses IEEE 802.16-2004, IEEE 802.16e-2005, IEEE 802.16-2009, and IEEE 802.16j-2009.

WiMAX wireless interface threats focus on compromising the radio links between WiMAX nodes. These radio links support both line-of-sight (LOS) and non-line-of-sight (NLOS) signal propagation. Links from LOS WiMAX systems are generally harder to attack than those from NLOS systems because an adversary would have to physically locate equipment between the transmitting nodes to compromise the confidentiality or integrity of the wireless link. WiMAX NLOS systems provide wireless coverage over large geographic regions, which expands the potential staging areas for both clients and adversaries. Like other networking technologies, all WiMAX systems must address threats arising from denial of service attacks, eavesdropping, man-in-the-middle attacks, message modification, and resource misappropriation.

To improve WiMAX system security, organizations should implement the following recommendations:

Organizations should develop a robust WMAN security policy and enforce it.

A security policy is an organization's foundation for designing, implementing, and maintaining properly secured technologies. WMAN policy should address the design and operation of the technical infrastructure and the behavior of users. Client devices should be configured to comply with WMAN policies, such as disabling unneeded services and altering default configurations. In addition, policy-driven software solutions can be implemented on client devices to prevent or allow certain actions to take place when specific conditions are met. Policy-driven software helps ensure that client devices and users comply with an organization's defined policies.

---

[1]   WiMAX was originally an acronym standing for Worldwide Interoperability for Microwave Access; however, WiMAX is no longer an acronym.

[2]   Information on the WiMAX Forum™ can be found at http://www.wimaxforum.org/home/.

Organizations should assess WiMAX technical countermeasures before implementing a vendor's WiMAX technology.

As of this writing, few WiMAX products employ Federal Information Processing Standard (FIPS) validated cryptographic modules. Consequently, vendors often integrate their WiMAX products with other security solutions that meet FIPS requirements. WiMAX interoperability certifications do not extend to these add-on approaches, which means there may be no assurance that the vendor's offering will function as intended. Given the diversity in potential approaches and the risk that integration issues could affect the security of the system, organizations should work closely with WiMAX vendors to gain a better understanding of potential system configuration constraints. Organizations should independently determine the need for compensating controls to address technical security functionality that the WiMAX product may not address.

Organizations using WiMAX technology should require mutual authentication for WiMAX devices.

WiMAX technology supports mutual device authentication between a base station (BS) and a user's subscriber unit (i.e., mobile phone, laptop, or similar device), but the feature must be activated to realize the benefit of the approach. Organizations should strongly consider WiMAX solutions capable of supporting Extensible Authentication Protocol (EAP) methods for mutual authentication as recommended in NIST SP 800-120, *Recommendation for EAP Methods Used in Wireless Network Access Authentication*.[3] EAP methods that support mutual device authentication typically also support integrated user authentication using passwords, smart cards, biometrics, or some combination of these mechanisms. WiMAX solutions that cannot meet these criteria should employ a different means of authentication at a higher layer (e.g., encryption overlay or virtual private network [VPN]). Specifically, native IEEE 802.16-2004 authentication does not support mutual device authentication and thus should be avoided.

Organizations using WiMAX networks should implement FIPS-validated encryption algorithms employing FIPS-validated cryptographic modules to protect data communications.

WiMAX communications consist of management and data messages. Management messages are used to govern communications parameters necessary to maintain wireless links, and data messages carry the data to be transmitted over wireless links. Encryption is not applied to management messages to increase the efficiency of network operations, while data messages are encrypted natively in accordance with the IEEE standards. IEEE 802.16e-2005 and IEEE 802.16-2009 support the Advanced Encryption Standard (AES) (as documented in FIPS Publication 197), whereas IEEE 802.16-2004 supports Data Encryption Standard in Cipher Block Chaining mode (DES-CBC). DES-CBC has several well-documented weaknesses, making it a vulnerable encryption algorithm that should not be used to protect data messages. Federal agency communications that require protection through encryption must use products with cryptographic functionality that is validated under the NIST Cryptographic Module Validation Program (CMVP), as meeting requirements per FIPS PUB 140. For WiMAX solutions that do not support FIPS-validated algorithms employing FIPS-validated cryptographic modules, organizations needing to protect the confidentiality of their WiMAX communications should deploy overlay encryption solutions, such as a FIPS-validated virtual private network solution.

---

[3]   NIST SP 800-120 can be found at http://csrc.nist.gov/publications/PubsSPs.html.

1. Introduction

1.1 Authority

The National Institute of Standards and Technology (NIST) developed this document in furtherance of its statutory responsibilities under the Federal Information Security Management Act (FISMA) of 2002, Public Law 107-347.

NIST is responsible for developing standards and guidelines, including minimum requirements, for providing adequate information security for all agency operations and assets; but such standards and guidelines do not apply to national security systems. This guideline is consistent with the requirements of the Office of Management and Budget (OMB) Circular A-130, Section 8b(3), "Securing Agency Information Systems," as analyzed in A-130, Appendix IV: Analysis of Key Sections. Supplemental information is provided in A-130, Appendix III.

This guideline has been prepared for use by Federal agencies. It may be used by nongovernmental organizations on a voluntary basis and is not subject to copyright although attribution is desired.

Nothing in this document should be taken to contradict standards and guidelines made mandatory and binding on Federal agencies by the Secretary of Commerce under statutory authority, nor should these guidelines be interpreted as altering or superseding the existing authorities of the Secretary of Commerce, Director of the OMB, or any other Federal official.

1.2 Purpose and Scope

The purpose of this document is to provide information to organizations regarding the security capabilities of wireless communications using WiMAX networks and to provide recommendations on using these capabilities. WiMAX technology is a wireless metropolitan area network (WMAN) technology based upon the IEEE 802.16 standard. It is used for a variety of purposes, including, but not limited to, fixed last-mile broadband access, long-range wireless backhaul, and access layer technology for mobile wireless subscribers operating on telecommunications networks.

The scope of this document is limited to the security of the WiMAX air interface and user subscriber devices, to include: security services for device and user authentication; data confidentiality; data integrity; and replay protection. This document does not address WiMAX network system specifications, which address core network infrastructure and are primarily employed by commercial network operators.[4] This publication, while containing requirements specific to Federal agencies, serves to provide security guidance for organizations considering the implementation of WiMAX systems.

1.3 Audience

This document discusses WiMAX technologies and security capabilities in technical detail. It assumes that the readers have at least some operating system, wireless networking, and security knowledge. Because of the constantly changing nature of the wireless security industry and the threats to and vulnerabilities of the technologies, readers are strongly encouraged to take advantage of other resources (including those listed in this document) for more current and detailed information.

---

[4] Security mechanisms, architecture, protocols, and procedures for WiMAX core infrastructure are specified in WiMAX Forum® Network Architecture Release 1.5 - Stage 3: Detailed Protocols and Procedure (http://www.wimaxforum.org/resources/documents/technical/T33).

The following list highlights people with differing roles and responsibilities that might use this document:

■ Government managers (e.g., chief information officers and senior managers) who oversee the use and security of WiMAX technologies within their organizations

■ Systems engineers and architects who design and implement WiMAX technologies

■ Auditors, security consultants, and others who perform security assessments of wireless environments

■ Researchers and analysts who want to understand the underlying wireless technologies.

1.4 Document Structure

The remainder of this document is composed of the following sections and appendices:

■ Section 2 reviews the technology components comprising the various WiMAX operating environments, the evolution of the IEEE 802.16 standard, and the WiMAX Forum product certifications.

■ Section 3 provides an overview of the security mechanisms included in the IEEE 802.16-2004, 802.16e-2005, 802.16-2009, and 802.16j-2009 specifications and highlights their limitations.

■ Section 4 examines common vulnerabilities and threats involving WiMAX technologies and makes recommendations for countermeasures to improve security.

■ Appendix A provides a glossary of key terms used in this document.

■ Appendix B consists of a list of acronyms and abbreviations used in this document.

■ Appendix C provides a list of references for this document.

2.   Overview of WiMAX Technology

A wireless metropolitan area network (WMAN) is a form of wireless networking that has an intended coverage area—a *range*—of approximately the size of a city. A WMAN is typically owned by a single entity such as an Internet service provider, government entity, or large corporation. Access to a WMAN is usually restricted to authorized users and subscriber devices.

The most widely deployed form of WMAN technology is WiMAX technology, which is based in large part on the IEEE 802.16 standard. The industry trade association, the WiMAX Forum[5], coined the WiMAX trademark and defines the precise content and scope of WiMAX technology through technical specifications that it creates and publishes. Early iterations of WiMAX technology (based on IEEE 802.16-2004 and earlier) were designed to provide fixed last-mile broadband wireless access. The IEEE 802.16e-2005 amendment added support for enhanced user mobility. The latest standard, IEEE 802.16-2009, consolidates IEEE 802.16-2004 and IEEE 802.16e-2005 in addition to IEEE 802.16 amendments approved between 2004 and 2008. IEEE also released IEEE 802.16j-2009 to specify multi-hop relay networking. This section explains the fundamental concepts of WiMAX technology, including its topologies, and discusses the evolution of the IEEE 802.16 standard.

2.1   Fundamental WiMAX Concepts

WiMAX networks have five fundamental architectural components:

- **Base Station (BS).** The BS is the node that logically connects wireless subscriber devices to operator networks. The BS maintains communications with subscriber devices and governs access to the operator networks. A BS consists of the infrastructure elements necessary to enable wireless communications, i.e., antennas, transceivers, and other electromagnetic wave transmitting equipment. BSs are typically fixed nodes, but they may also be used as part of mobile solutions—for example, a BS may be affixed to a vehicle to provide communications for nearby WiMAX devices. A BS also serves as a Master Relay-Base Station in the multi-hop relay topology (described in Section 2.2).

- **Subscriber Station (SS).** The SS is a stationary WiMAX-capable radio system that communicates with a base station, although it may also connect to a relay station in multi-hop relay network operations (described in Section 2.2).

- **Mobile Station (MS).** An MS is an SS that is intended to be used while in motion at up to vehicular speeds. Compared with fixed (stationary) SSs, MSs typically are battery operated and therefore employ enhanced power management. Example MSs include WiMAX radios embedded in laptops and mobile phones. This document uses the term SS/MS to refer to the class of both MS and stationary SS.[6]

- **Relay Station (RS).** RSs are SSs configured to forward traffic to other RSs or SSs in a multi-hop Security Zone (which is discussed in Section 3.5). The RS may be in a fixed location (e.g., attached to a building) or mobile (e.g., placed in an automobile). The air interface between an RS and an SS is identical to the air interface between a BS and an SS.

- **Operator Network** – The operator network encompasses infrastructure network functions that provide radio access and IP connectivity services to WiMAX subscribers. These functions are defined in WiMAX Forum technical specifications as the access service network (radio access) and the

---

5   Information on the WiMAX Forum can be found at http://www.wimaxforum.org/home/.
6   In a strict reading of the IEEE 802.16 standards, an SS can refer to both an MS and a stationary (location-fixed) SS. However, many WiMAX documents refer to SS as fixed stations only. The term SS/MS clarifies this ambiguity by explicitly indicating that it refers to either fixed or mobile stations.

connectivity service network (IP connectivity).[7] WiMAX devices communicate using two wireless message types: management messages and data messages. *Data messages* transport data across the WiMAX network. *Management messages* are used to maintain communications between an SS/MS and BS, e.g., establishing communication parameters, exchanging security settings, and performing system registration events (initial network entry, handoffs, etc.)

IEEE 802.16 defines frequency bands for operations based on signal propagation type. In one type, it employs a radio frequency (RF) beam to propagate signals between nodes. Propagation over this beam is highly sensitive to RF obstacles, so an unobstructed view between nodes is needed. This type of signal propagation, called *line-of-sight (LOS)*, is limited to fixed operations and uses the 10 to 66 gigahertz (GHz) frequency range. The other type of signal propagation is called *non-line-of-sight (NLOS)*. NLOS employs advanced RF modulation techniques to compensate for RF signal changes caused by obstacles that would prevent LOS communications. NLOS can be used for both fixed WiMAX operations (below 11 GHz) and mobile operations (below 6 GHz). NLOS signal propagation is more commonly employed than LOS because of obstacles that interfere with LOS communications and because of strict regulations for frequency licensing and antenna deployment in many environments that hinder the feasibility of using LOS.

2.2 Operating Topologies

There are four primary topologies for IEEE 802.16 networks: point-to-point, point-to-multipoint, multi-hop relay, and mobile. Each of these topologies is briefly described below.

2.2.1 Point-to-Point (P2P)

A *point-to-point (P2P) topology* consists of a dedicated long-range, high-capacity wireless link between two sites. Typically, the main or central site hosts the BS, and the remote site hosts the SS, as seen in Figure 2-1. The BS controls the communications and security parameters for establishing the link with the SS. The P2P topology is used for high-bandwidth wireless backhaul[8] services at a maximum operating range of approximately 48 kilometers (km) (30 miles) using LOS signal propagation, and eight km (five miles) using NLOS.

---

[7]   WiMAX Forum specifications can be found at http://www.wimaxforum.org/resources/documents/technical/.

[8]   A backhaul is typically a high capacity line from a remote site or network to a central site or network.

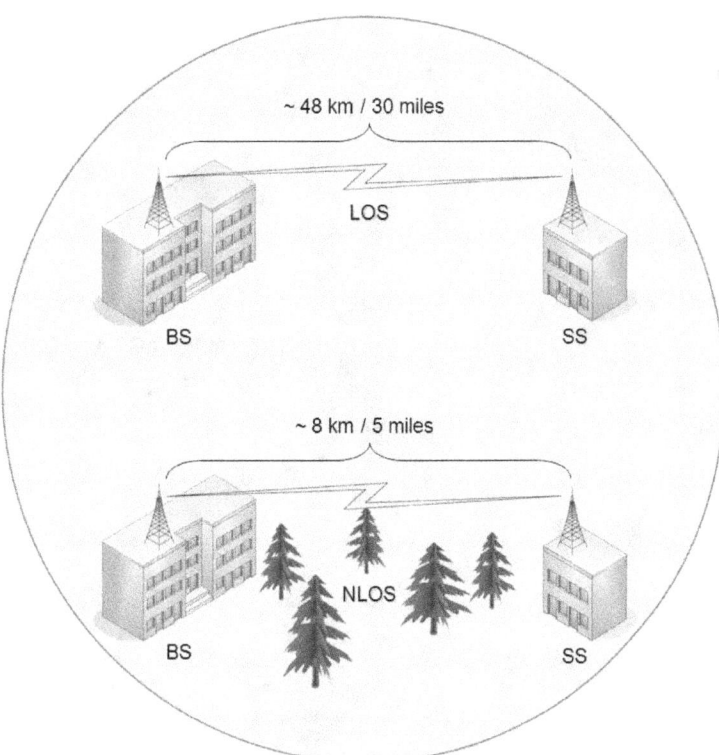

Figure 2-1. P2P Topology

2.2.2 Point-to-Multipoint (PMP)

A *point-to-multipoint (PMP) topology* is composed of a central BS supporting multiple SSs, providing network access from one location to many. It is commonly used for last-mile broadband access,[9] private enterprise connectivity to remote offices, and long-range wireless backhaul services for multiple sites. PMP networks can operate using LOS or NLOS signal propagation. Each PMP BS has a maximum operating range of 8 km (5 miles), but it is typically less than this due to cell configuration and the urban density of the target coverage area. Figure 2-2 illustrates the PMP topology.

---

[9]   Last-mile broadband access refers to communications technology that bridges the transmission distance between the broadband service provider infrastructure and the customer premises equipment.

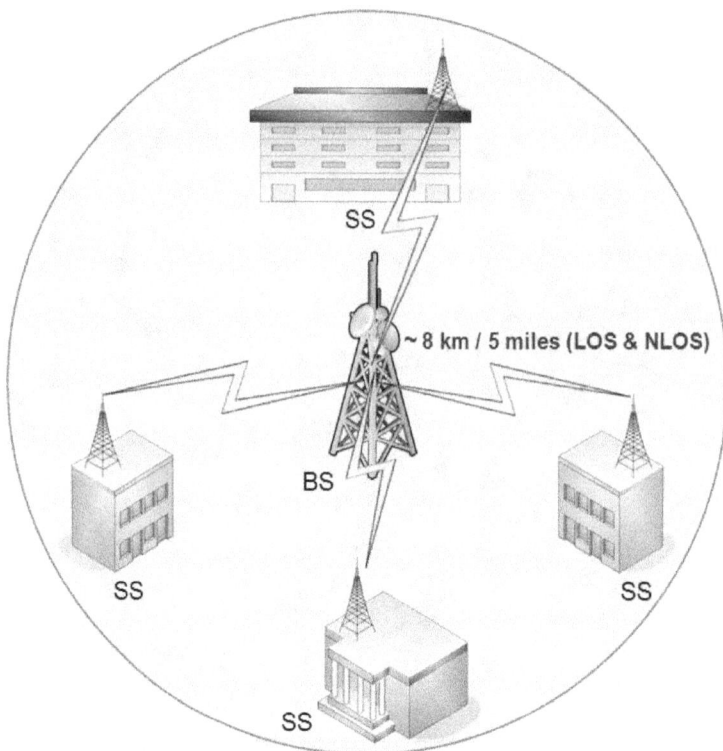

Figure 2-2. PMP Topology

2.2.3 Multi-Hop Relay

A *multi-hop relay topology,* defined by IEEE 802.16j-2009, extends a BS's coverage area by permitting SSs/MSs to relay traffic by acting as RSs. Data destined to an SS/MS outside of the BS's range is relayed through adjacent RSs. An RS can only forward traffic to RSs/SSs within its Security Zone. A *Security Zone* is a set of trusted relationships between a BS and a group of RSs. Data originating outside of a BS's coverage area is routed over multiple RSs, increasing the network's total geographical coverage area, as seen in Figure 2-3. Multi-hop relay topology typically uses NLOS signal propagation because its purpose is to span large geographic areas containing multiple RF obstacles; however, technically it can operate using LOS propagation as well. The maximum operating range for each node in a multi-hop relay topology is approximately 8 km (5 miles), but the actual operating range is typically less depending on environmental conditions (e.g., building obstructions) and antenna configuration.

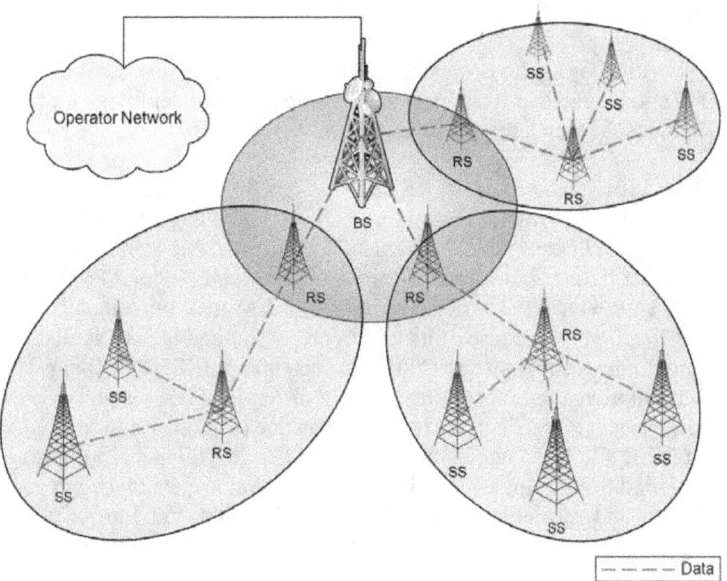

Figure 2-3. Multi-Hop Topology

2.2.4 Mobile

A *mobile topology* is similar to a cellular network because multiple BSs collaborate to provide seamless communications over a distributed network to both SSs and MSs. This topology combines the coverage area of each member BS and includes measures to facilitate handoffs of MSs between BS coverage areas, as seen by the car MS in Figure 2-4. It uses advanced RF signaling technology to support the increased RF complexity required for mobile operations. Each BS coverage area is approximately 8 km (5 miles). Mobile WiMAX systems operate using NLOS signal propagation on frequencies below 6 GHz.

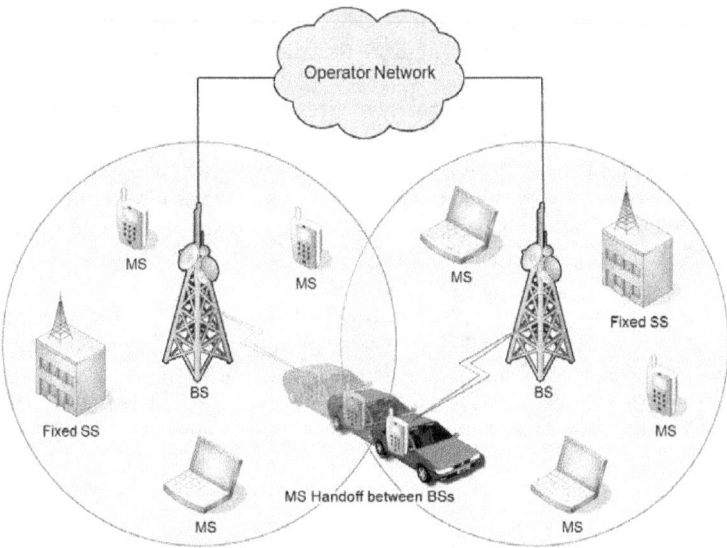

Figure 2-4. Mobile Topology

2.3   Evolution of the IEEE 802.16 Standard

In 1999, the 802.16 Working Group on Broadband Wireless Access Standards was established to develop standards and recommended practices to support the development and deployment of broadband WMAN. The following summarizes the standards and amendments that the working group has produced.

When the IEEE 802.16-2001 standard, which targeted last-mile broadband wireless access, was first under development in 2001, the WiMAX Forum formed to promote the compatibility and interoperability of IEEE 802.16 technologies. In December 2001, the IEEE 802.16-2001 standard was approved. It operated in the 10–66 GHz frequency range and provided LOS fixed P2P and PMP communications at maximum data rates of approximately 70 megabits per second (Mbps). Implementation of the IEEE 802.16-2001 standard was limited because of its LOS requirement and lack of available spectrum. Because of the shortcomings of IEEE 802.16-2001, an amendment, IEEE 802.16a, was released during 2003. This amendment improved interoperability, quality of service (QoS), and data performance. It also provided the ability to propagate signals from one active device to another and to have NLOS communications. An IEEE 802.16d amendment was also under development to improve interoperability, but it was later transitioned from a single amendment to a revision project that aggregated IEEE 802.16-2001 and its amendments under a single standard: IEEE 802.16-2004. Versions of IEEE 802.16 prior to IEEE 802.16-2004 are no longer supported by vendors and are not discussed further in this publication.

IEEE 802.16-2004 combines all of the improved functionality of the IEEE 802.16-2001 amendments with new interoperability specifications. The standard supports communications in the 2–66 GHz frequency range, with 10–66 GHz for LOS and 2–11 GHz for NLOS. Each frequency range employs different modulation techniques to accommodate LOS and NLOS communications requirements. Additionally, IEEE 802.16-2004 can operate in P2P and PMP topologies.

The IEEE 802.16e-2005 amendment to the IEEE 802.16-2004 standard provides enhancements to fixed wireless operations and enables a cellular-like architecture. Specifically, IEEE 802.16e-2005 provides mobility support for SSs and implements enhanced signaling techniques that enable new service offerings such as Voice over Internet Protocol (VoIP), presence, and multimedia broadcast. These enhancements to the previous standard's QoS make it resilient against communications latency and jitter.[10] Additionally, IEEE 802.16e-2005 limits the frequency range to 6 GHz or below for mobile operations. IEEE 802.16e-2005 also introduces new security measures; these are described in detail in Section 3.

Providing support in IEEE 802.16e-2005 for mobile devices necessitated a significant departure from the process IEEE 802.16-2004 BSs use to manage SSs. IEEE 802.16e-2005 introduces dynamic roaming and other new methods to manage the communication handoffs between SSs and BSs, i.e., switching an SS transmission from one BS coverage area into a new BS coverage area as the mobile SS moves. The communications architecture is also modified to facilitate better power management and efficient modes of operation to address the power constraints of MSs.

In May 2009, IEEE consolidated 802.16-2004, 802.16e-2005, 802.16f-2005, and 802.16g-2007 into the latest IEEE 802.16-2009 standard. IEEE 802.16-2009 technically made the consolidated standards and amendments obsolete; however, as of this writing, many production WiMAX networks are still based on IEEE 802.16-2004 or IEEE 802.16e-2005. In June 2009, IEEE released the IEEE 802.16j-2009 amendment specifying multi-hop relay. This amendment provides a more developed and thorough

---

[10]   Jitter, as it relates to queuing, is the difference in latency of packets.

security and communications architecture for multi-hop networking than was previously defined in IEEE 802.16-2004 for the mesh networking option.[11]

In addition to the standards and amendments already discussed, Table 2-1 lists other relevant standards and amendments to the IEEE 802.16 family of standards.

Table 2-1. Additional IEEE 802.16 Standards and Amendments

Name	Standard or Amendment	Status[12]	Purpose
802.16h	Amendment	Current Draft: 3/2010	Develops methods to improve WiMAX coexistence over license-exempt spectrum.
802.16k-2007	Standard	Active: Published 8/2007	Defines procedures to support bridge functionality in IEEE 802.16-2004.
802.16m	Standard	Current Draft: 4/2010	Enhances the IEEE 802.16 air interface to support speeds up to 1 gigabit/second (Gbps) for fixed operations and 100 Mbps for mobile operations.

Generally, standards bodies provide a framework for product development but cannot ensure vendor interoperability. Product certifications serve to encourage market adoption of standards-based technology and to validate vendor operability claims. A major effort of the WiMAX Forum is designing and promoting the certification of products based on the IEEE 802.16 standard. The WiMAX Forum operates the WiMAX Forum Certification Program using accredited testing laboratories and designated certification bodies to ensure WiMAX products are compatible, interoperable, and conform to industry standards to ensure the interoperability of different vendor products. This results in greater competition in the marketplace, greater flexibility in deployment, larger target markets, and lower production costs.[13] The WiMAX Forum warns that "vendors claiming their equipment is 'WiMAX-like', 'WiMAX-compliant', and etcetera are not WiMAX Forum Certified, which means that their equipment may not be interoperable with other vendors' equipment."[14]

---

[11] The lack of a robust mesh networking security and communications architecture in IEEE 802.16-2004 resulted in low market adoption of IEEE 802.16-2004 based mesh networking.

[12] *Active* indicates that the standard or amendment has been approved by the IEEE standards board, and *Draft* indicates that it is awaiting tentative approval in Sponsor Ballot.

[13] The WiMAX Forum Certified Product Registry can be found at http://www.wimaxforum.org/productshowcase.

[14] WiMAX Forum, "Certification Program," http://www.wimaxforum.org/certification/program

3.   WiMAX Security Features

This section discusses the security mechanisms included in IEEE 802.16-2004, IEEE 802.16e-2005, IEEE 802.16-2009, and IEEE 802.16j-2009[15]; it illustrates their functions and provides a foundation for the security recommendations in Section 4. The IEEE 802.16 standards specify two basic security services: authentication and confidentiality. Authentication involves the process of verifying the identity claimed by a WiMAX device. Confidentiality is limited to protecting the contents of WiMAX data messages so that only authorized devices can view them. IEEE 802.16e-2005 and IEEE 802.16-2009 share the same authentication and confidentiality mechanisms; they both support user authentication and device authentication.

The IEEE 802.16 standards do not address other security services such as availability and confidentiality protection for wireless management messages [16]; if such services are required, they must be provided through additional means. Also, while IEEE 802.16 security protects communications over the WMAN link between an SS/MS and a BS, it does not protect communications on the wired operator network behind the BS. End-to-end (i.e., device-to-device) security is not possible without applying additional security controls not specified by the IEEE standards.

WiMAX systems provide secure communications by performing three steps: authentication, key establishment, and data encryption. Figure 3-1 is a high-level overview of the security framework. The authentication procedure provides common keying material for the SS/MS and the BS and facilitates the secure exchange of data encryption keys that ensure the confidentiality of WiMAX data communications. The remainder of this section explains the basics of the WiMAX security framework, authentication, key establishment, and data encryption[17].

---

[15]   IEEE 802.16j-2009 contains all of the security capabilities of IEEE 802.16-2009, and also includes security features for multi-hop relay networking. All discussions of IEEE 802.16-2009 security in this publication are also applicable to IEEE 802.16j-2009.

[16]   Management messages are typically not encrypted in wireless communications because adding encryption could negatively affect network operations. A common example of a management message is a ranging request used to determine network delay at initial network entry and periodically during operations. Because this message is time sensitive, encrypting it would diminish network availability.

[17]   In Figure 3-1, "Data Encryption" represents not only data encryption, but also control message authentication, integrity, and replay protection.

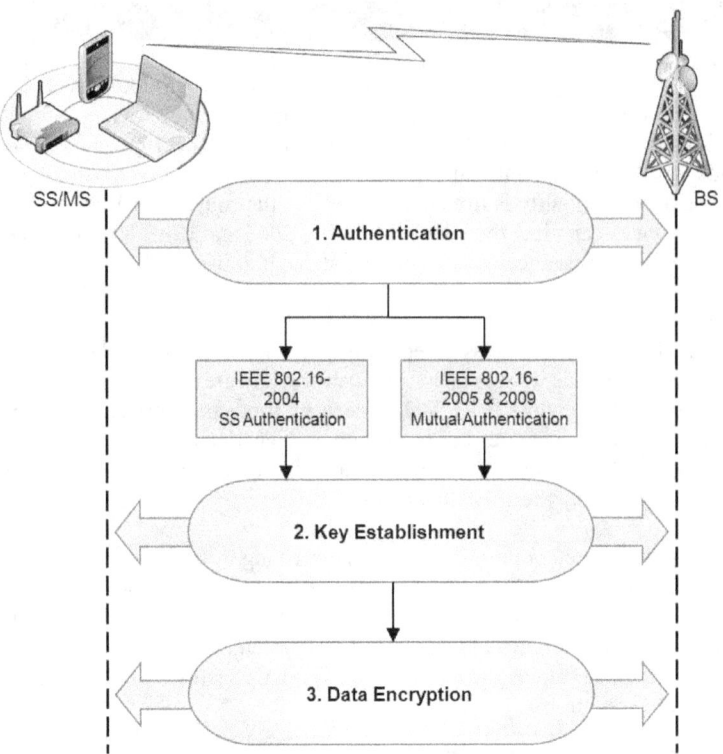

Figure 3-1. WiMAX Security Framework

3.1 Security Associations

A security association (SA) is a shared set of security parameters that a BS and its SS/MS use to facilitate secure communications. Similar in concept to Internet Protocol Security (IPsec),[18] an SA defines the security parameters of a connection, i.e., encryption keys and algorithms. SAs fall into one of three categories: authorization, data (for unicast[19] services), and group (for multicast[20] services). A distinct SA is established for each service offered by the BS. For example, a unicast service would have a unique data encryption SA, whereas a multicast service would have a unique group SA.

Authorization SAs facilitate authentication and key establishment to configure data and group SAs. Authorization SAs contain the following attributes:

- **X.509 certificates.** X.509 digital certificates allow WiMAX communication components to validate one another. The manufacturer's certificate is used for informational purposes, and the BS and SS/MS certificates contain the respective devices' public keys. The certificates are signed by the device manufacturer or a third-party certification authority.

- **Authorization key (AK).** AKs are exchanged between the BS and SS/MS to authenticate one another prior to the traffic encryption key (TEK) exchange. The authorization SA includes an identifier and a key lifetime value for each AK.

---

[18] IPsec is a network layer security protocol that provides authentication and encryption over untrusted networks.
[19] Unicast traffic is a single data transmission sent to a single recipient over the same network.
[20] Multicast traffic is a single data transmission sent to multiple recipients over the same network.

■ **Key encryption key (KEK).** Derived from the AK, the KEK is used to encrypt TEKs during the TEK exchange, as discussed in Section 3.3.

■ **Message authentication keys.** Derived from the AK, the message authentication keys validate the authenticity of key distribution messages during key establishment. These keys are also used to sign management messages to validate message authenticity.

■ **Authorized data SA list.** Provided to the SS/MS by the BS, the authorized data SA list indicates which data encryption SAs the SS/MS is authorized to access.

Data SAs establish the parameters used to protect unicast data messages between BSs and SSs/MSs. Data SAs cannot be applied to management messages, which are never encrypted. A data SA contains the following security attributes:

■ **SA identifier (SAID).** This unique 16-bit value identifies the SA to distinguish it from other SAs.

■ **Encryption cipher to be employed.** The connection will use this encryption cipher definition to provide wireless link confidentiality.

■ **Traffic encryption key (TEK).** TEKs are randomly generated by the BS and are used to encrypt WiMAX data messages. Two TEKs are issued to prevent communications disruption during TEK rekeying; the first TEK is used for active communications, while the second TEK remains dormant.[21]

■ **Data encryption SA type indicator.** This indicator identifies the type of data SA. There are three types:

  – Primary SA. This SA is established as a unique connection for each SS/MS upon initialization with the BS. There is only one primary SA per SS/MS.

  – Static SA. This SA secures the data messages and is generated for each service defined by the BS.

  – Dynamic SA. This SA is created and eliminated in response to the initiation and termination of specific service flows.

Group SAs contain the keying material used to secure multicast traffic. Group SAs are inherently less secure than data SAs because identical keying material is shared among all members of a BS's group. Group SAs contains the following attributes:

■ **Group traffic encryption key (GTEK).** This key is randomly generated by the BS and used to encrypt multicast traffic between a BS and SSs/MSs.

■ **Group key encryption key (GKEK).** This key is also randomly generated by the BS and used to encrypt the GTEK sent in multicast messages between a BS and SSs/MSs.

3.2 Authentication and Authorization

Networking technologies traditionally refer to authorization as the process that determines the level of access a node receives after the subject is identified and authenticated. The IEEE 802.16 standard generally refers to authorization as the process of authenticating WiMAX nodes and granting them access to the network. This slight distinction made by IEEE 802.16 is that authorization processes implicitly include authentication. The Privacy Key Management (PKM) protocol is the set of rules responsible for

---

[21] Per the IEEE 802.16-2004 standard, "Each TEK becomes active halfway through the lifetime of its predecessor and expires halfway through the lifetime of its successor" [IEE04]. This cyclical process ensures that keying material is continually refreshed.

authentication and authorization to facilitate secure key distribution in WiMAX. PKM uses authorization SAs to authenticate system entities so that data and group encryption SAs can be established. PKM's authentication enforcement function provides the SS/MS and BS with identical AKs; each AK is then used to derive the message authentication keys and KEKs that facilitate the secure exchange of the TEKs. IEEE 802.16-2004 derives the AK using PKM version 1 (PKMv1), whereas IEEE 802.16e-2005 and IEEE 802.16-2009 derive the AK using PKMv2. This section reviews the procedures used in both PKMv1 and PKMv2.

3.2.1   IEEE 802.16-2004 Authentication and Authorization

In PKMv1 [IEE04], the BS authenticates the identity of the SS, providing one-way authentication. Figure 3-2 illustrates the challenge-response verification scheme used in PKMv1-based authentication. The authorization process is initiated when the SS sends an authorization information message to the BS. This message contains the X.509 certificate of the SS manufacturer and is used by the BS for informational purposes. Immediately following the authorization information message, the SS sends an authorization request to the BS, which contains the following information:

- The SS's unique X.509 certificate, which includes its RSA public key

- A description of the SS's supported cryptographic algorithms

- The primary SAID

Next, the BS validates the SS's X.509 certificate, communicates the supported cryptographic algorithms and protocols, and activates an AK for the SS. Then the BS sends the SS an authorization reply message containing the following information:

- The activated AK, encrypted with the SS's public key

- The AK sequence number used to differentiate between successive generations of AKs

- The AK lifetime

- A list of SAIDs that the SS is authorized to access and their associated properties

The AK is periodically reauthorized based on its lifetime. The reauthorization process is identical to the initial authorization process with the exception that the authorization information message is not re-sent. Reauthorization does not cause a service interruption because two AKs with overlapping lifetimes are supported simultaneously.

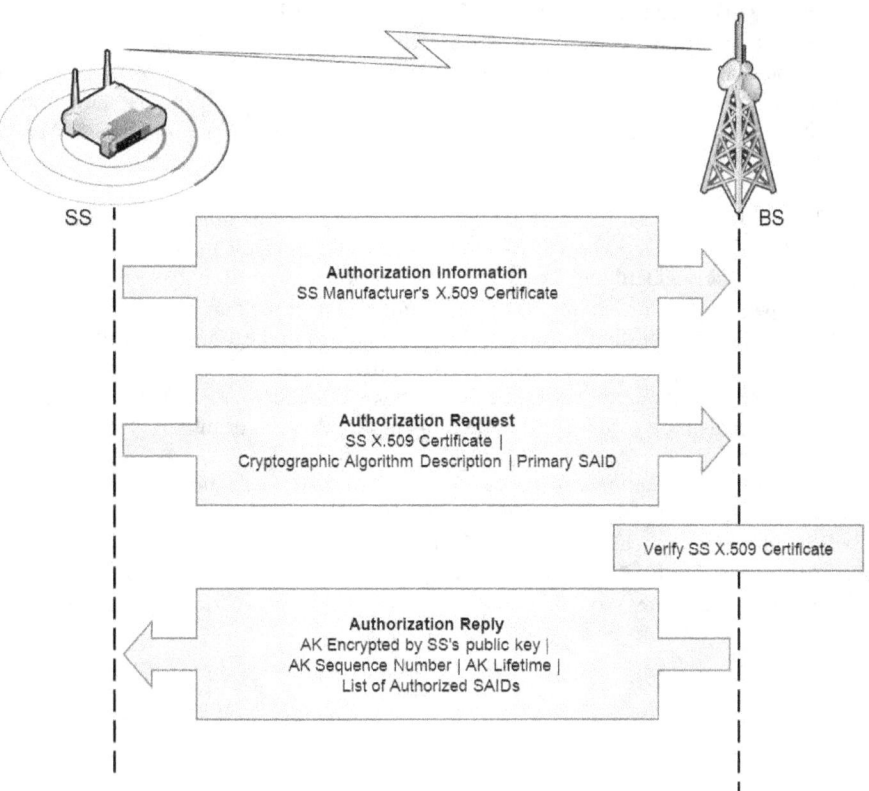

Figure 3-2. PKMv1 Authorization

3.2.2   IEEE 802.16-2009 and WiMAX Forum Network Architecture Release 1.5

IEEE 802.16-2009 includes security features of the 802.16e-2005 amendment, which was adopted after the publication of 802.16-2004. The WiMAX Forum Network Architecture Release 1.5 further extends the security framework. In particular, the Base Specification delineates the required Extensible Authentication Protocol (EAP) methods that a certified device must support, and describes the use of Remote Authentication Dial-In User Services (RADIUS) (and its Diameter successor) for authentication, authorization, and accounting (AAA). The addition of EAP and RADIUS/Diameter support enables WiMAX networks to be tailored to a wide range of robust enterprise security architectures, and also makes the design and implementation of WiMAX networks more complex than had been the case with IEEE 802.16-2004.

The WiMAX Forum Network Architecture Release 1.5 states requirements for device and user authentication. For mutual device authentication based on X.509 certificates, an SS/MS must support EAP-transport layer security (EAP-TLS). For user authentication, the SS/MS must support either EAP-authentication and key agreement (EAP-AKA) or EAP-tunneled TLS (EAP-TTLS), preferably both. EAP-AKA is an authentication method used in Universal Mobile Telecommunications System (UMTS) and CDMA2000 networks. It is based on symmetric key encryption that typically runs in a subscriber identity module (SIM) or similar smart card.[22] EAP-TTLS authenticates the network to the user with an X.509 certificate and authenticates the user to the network with another "tunneled" EAP method. The

---

[22]   EAP-AKA is described in RFC 4187, which can be found at http://tools.ietf.org/html/rfc4187.

WiMAX Forum Network Architecture Release 1.5 requires that EAP-TTLS support Microsoft Challenge-handshake authentication protocol version 2 (MS-CHAPv2) at a minimum. Vendors may implement other EAP methods at their discretion to support specialized authentication requirements. Organizations should strongly consider WiMAX solutions capable of supporting EAP methods for mutual authentication as recommended in NIST SP 800-120, *Recommendation for EAP Methods Used in Wireless Network Access Authentication.*

IEEE 802.16-2009 also specifies a Rivest, Shamir, Adleman (RSA) authentication protocol for mutual device authentication that uses X.509 certificates that contain the device's media access control (MAC) address. According to the standard, devices that use this protocol must either have factory-installed public/private key pairs or provide an internal algorithm to generate the pair automatically. The method has no known security vulnerabilities. However, it is not included in the WiMAX Forum Network Architecture Release 1.5 and, consequently, WiMAX certified products do not necessarily have an RSA Authentication feature. Additionally, IEEE 802.16-2009 RSA Authentication should not be confused with EAP methods that also use X.509 certificates and employ RSA algorithms.

Figure 3-3 depicts the *EAP authentication* procedure. The first EAP exchange results in the production of a 512-bit master session key (MSK) that is disclosed to the AAA server, the operator network, and the SS/MS. The BS and SS/MS truncate the MSK to 320 bits – 160 bits for the pairwise master key (PMK) and 160 bits to create another EAP Integrity Key (EIK) to protect an optional EAP user authentication procedure. The PMK, the SS/MS MAC address, and the BS identifier are then used to derive the AK.

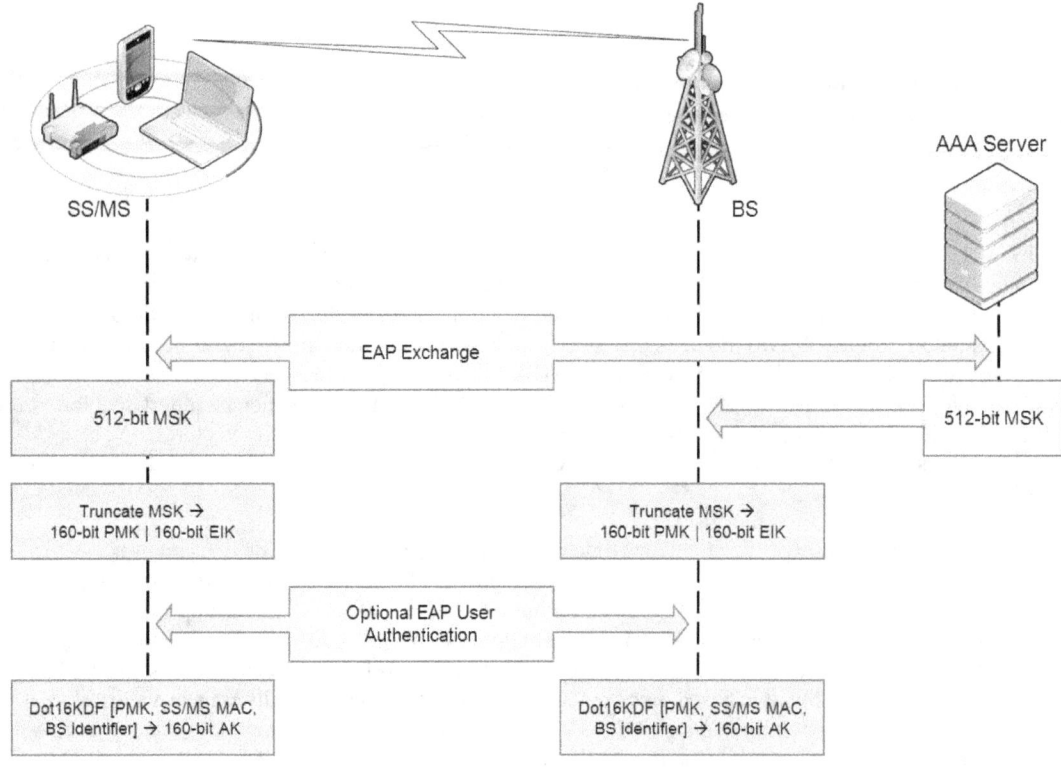

Figure 3-3. EAP Authentication

3.3 Encryption Key Establishment

Once authentication is complete, the BS and SS/MS share an activated AK. PKM then uses the 160-bit AK to derive the 128-bit KEK and the 160-bit message authentication keys, which are used to facilitate a secure exchange of TEKs.[23] The secure TEK exchange uses a three-way handshake between the BS and the SS/MS, as illustrated in Figure 3-4.[24]

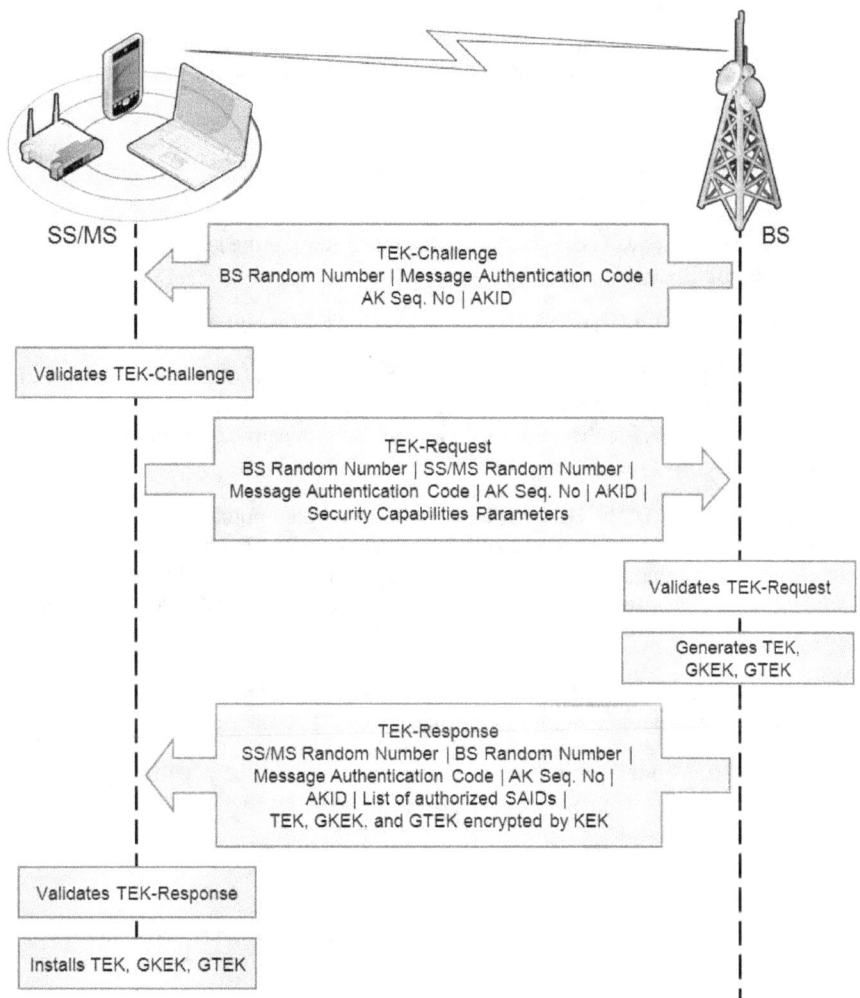

Figure 3-4. TEK Three-Way Handshake

The first step in this procedure is the TEK-Challenge sent from the BS to the SS/MS. The TEK-Challenge is sent during initial network entry or during reauthorization. The TEK-Challenge includes the following attributes:

---

[23] In some cases, the message authentication keys can be derived from the EIK during the EAP exchange.

[24] The depiction of the handshake in Figure 3-6 illustrates a general framework for TEK exchange defined in PKM. TEK exchanges will vary slightly depending on which encryption algorithm is being used.

- **BS random number.** This number is attached to the TEK-Challenge to prevent replay attacks[25] by validating message freshness.

- **Message authentication code.** This validates data authenticity of the key distribution messages sent from the BS to the SS/MS.

- **AK sequence number and AK identifier (AKID).** These attributes identify which AK is used for the TEK exchange.

Upon receipt of the TEK-Challenge, the SS/MS validates the authenticity of the TEK-Challenge using the message authentication keys. After the TEK-Challenge has been validated, the SS/MS then sends the TEK-Request to the BS, which contains the following attributes:

- **BS and SS/MS random numbers.** In addition to sending back the BS random number from the TEK-Challenge, the SS/MS attaches its own random value.

- **Message authentication code.** These validate data authenticity of the key distribution messages sent from the SS/MS to the BS.

- **AK sequence number and AKID.** These identify which AK is used for the TEK exchange.

- **Security capabilities parameters.** These describe the security capabilities of the SS/MS, including supported cryptographic suites. During initial network entry, the TEK-Request will also include a request for SA descriptors to identify the primary, static, and dynamic SAs that the SS/MS is authorized to access.

Upon receipt of the TEK-Request, the BS verifies that the BS random number matches the number sent in the TEK-Challenge and validates the message authentication keys. The BS next confirms that the AKID refers to an available AK and that the security capabilities parameters provided by the SS/MS are supported. Once the TEK-Request is validated, the BS will generate two TEKs, along with the GKEK and the GTEK. The BS then sends the TEK-Response to the SS/MS, which contains the following attributes:

- **BS and SS/MS random number.** The BS attaches the BS random number generated in the TEK-Challenge and the SS/MS random number generated in the TEK-Request.

- **Message authentication code.** These validate data authenticity for the key distribution messages sent from the BS to the SS/MS.

- **AK sequence number and AKID.** These attributes identify which AK is used for the TEK exchange.

- **List of authorized SAIDs.** This is the list of primary, static, and dynamic SAs that the SS/MS is authorized to access.

- **TEKs, GKEK, and GTEK.** Using the KEK derived from the AK, the BS encrypts the two TEKs, the GKEK, and the GTEK. These keys include all of the required keying material needed to facilitate secure communications.

Upon receipt of the TEK-Response, the SS/MS will ensure the BS random number matches the value given in the TEK-Challenge and that the SS/MS random number matches the value delivered in the TEK-

---

[25] Replay attacks involve capturing legitimate traffic and then replaying it at a later time for malicious purposes. A standard replay attack involves replaying authentication credentials to gain unauthorized system access. Mechanisms for protecting against replay attacks typically involve inserting data that unpredictably varies over time, thereby ensuring that messages captured at a given point in time will not be useful sometime thereafter.

Request. The SS/MS will then validate the message authentication keys. Once validation is complete, the SS/MS will install the appropriate TEKs, GTEK, and GKEK, and secure communications can begin.

In the case of an MS performing a handover to a new BS, the TEK-Response message also includes TEK, GTEK, and GKEK parameters of the previously serving BS to reduce latency associated with renewing SAs.

3.4   Data Confidentiality

The completion of the TEK exchange provides the SS/MS and BS with the TEKs required to encrypt WiMAX data communications. The type of encryption employed by the TEK varies by IEEE 802.16 standard.

IEEE 802.16-2004 only supports one encryption algorithm, the Data Encryption Standard (DES) in cipher block chaining (CBC) mode (DES-CBC). During the TEK three-way handshake, the BS sends the SS an SA-specific initialization vector (IV) as part of the TEK-Response. The DES-CBC algorithm uses this SA-specific IV in conjunction with the TEK to encrypt data traffic. DES-CBC has significant weaknesses and should not be used to provide confidentiality for communications.[26]

IEEE 802.16e-2005 and IEEE 802.16-2009 support DES-CBC and three AES[27] modes of operation for data encryption: CBC, counter (CTR), and CTR with CBC message authentication code (CCM). Any of the three specified AES modes is acceptable for protecting data message confidentiality.[28] CTR mode is considered stronger than CBC because CTR mode is less complex to implement, offers encryption block preprocessing, and can process data in parallel. CCM mode enhances CTR by adding the capability to verify the authenticity of encrypted messages. CCM is considered the most secure of the cryptographic suites defined in IEEE 802.16e-2005 and IEEE 802.16-2009 because it adds a per packet randomization integrity check that prevents replay attacks.[29] Because of this, whenever feasible the CCM mode should be used instead of CTR or CBC.

CCM was specifically designed to have the following characteristics [Fra07]:

■   A single cryptographic key for confidentiality and integrity to minimize complexity and maximize performance (minimize key scheduling time)

■   Integrity protection of the packet header and packet payload, in addition to providing confidentiality of the payload

■   Computation of some cryptographic parameters prior to the receipt of packets to enable fast comparisons when they arrive, which reduces latency

■   Small footprint (hardware or software implementation size) to minimize costs

■   Small security-related packet overhead (e.g., minimal data expansion due to cryptographic padding and integrity field)

---

[26]   Additionally, DES-CBC is not approved for Federal agency use in protecting the confidentiality of communications.

[27]   AES is defined by FIPS PUB 197 (http://csrc.nist.gov/publications/fips/fips197/fips-197.pdf).

[28]   NIST SP 800-38A, *Recommendation for Block Cipher Modes of Operation*, recommends the use of five confidentiality modes of operation for symmetric key block cipher algorithms. The modes may be used in conjunction with any symmetric key block cipher algorithm that is approved by a FIPS. The five modes—the Electronic Codebook (ECB), CBC, Cipher Feedback (CFB), Output Feedback (OFB), and CTR modes—can provide data confidentiality.

[29]   CCM is defined by RFC 3610, *Counter with CBC-MAC (CCM)*, http://www.ietf.org/rfc/rfc3610.txt.

Additional information on data encryption, particularly Federal agency-specific requirements, is presented in Section 4.3.3.

3.5 IEEE 802.16j-2009 Multi-Hop Relay Security Architecture

The confidentiality and authentication security mechanisms used in IEEE 802.16j-2009 are identical to those in IEEE 802.16-2009. An additional security mechanism is required to operate a WiMAX network in a multi-hop relay—the establishment of a Security Zone (SZ). An SZ is the set of trusted relationships between a BS (acting as the master relay), RSs, and SSs/MSs. RSs and SSs/MSs become members of a BS's SZ by authenticating using PKMv2. Upon authenticating, the BS delivers SZ key material used to provide integrity protection to management messages in the SZ.

Multi-hop relay can operate in two security control architectures, centralized and distributed. In a centralized security architecture, RSs forward data and management messages destined for the next hop in the network without performing any decryption or authentication of management messages. Thus, every SA is established between each node and the BS with RSs strictly relaying traffic.

The IEEE 802.16j-2009 distributed security architecture provides greater scalability because network nodes are only required to establish SAs to adjacent nodes. For example, imagine an SS/MS that requires two RS hops to communicate back to the master relay BS. To create a secure route in a distributed architecture, the initial RS hop, referred to as the *access RS*, establishes an SA to the master relay BS. Then, the access RS relays authentication messages on behalf of the master relay BS to the next hop or *subordinate RS*. Upon establishing the MSK between the BS and the subordinate RS, the BS will forward the relevant AK to the first hop access RS. This allows the access RS to derive all of the necessary keys to establish a separate SA with the subordinate RS. The subordinate RS then establishes a separate SA with the destination SS/MS by employing the same method it used to establish the SA with the access RS. The culmination of these distributed SAs can be seen in Figure 3-5. There is now a secure tunnel between the BS and destination SS/MS while simultaneously load-balancing encryption processing away from the master BS. The logical security risk associated with a distributed model that shares keying information across multiple nodes is negligible because all SAs are established using PKMv2.

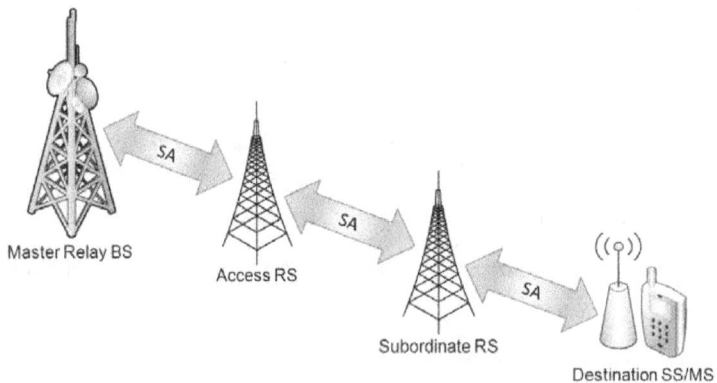

Figure 3-5. Multi-Hop Distributed Security Architecture

4. Vulnerabilities, Threats, and Countermeasures for WiMAX Systems

This section describes vulnerabilities in the IEEE 802.16 specifications, threats to WiMAX systems, and countermeasures to mitigate those threats. WiMAX systems are susceptible to WiMAX-specific threats as well as threats common to all wireless technologies. Organizations should mitigate threats and vulnerabilities for WiMAX systems by implementing a combination of management, operational, and technical countermeasures.[30]

4.1 Vulnerabilities

This section discusses vulnerabilities specific to IEEE 802.16-2004, IEEE 802.16e-2005, and IEEE 802.16-2009. (All discussion related to IEEE 802.16-2009 also applies to IEEE 802.16j-2009.) IEEE 802.16-2004 has more innate vulnerabilities than IEEE 802.16e-2005 or IEEE 802.16-2009. The security mechanisms available in IEEE 802.16e-2005 and IEEE 802.16-2009 address some, but not all, of IEEE 802.16-2004's shortcomings. The following discusses several major vulnerabilities:

- **Lack of BS to SS/MS authentication.** PKMv1 defines authentication of SSs by BSs but provides no means to authenticate BSs by SSs/MSs. Lack of mutual authentication may allow a rogue BS to impersonate a legitimate BS, thereby rendering the SS/MS unable to verify the authenticity of protocol messages received from the BS. A successful attack would enable a rogue BS operator to take complete control of all traffic to and from the SS/MS, including capture of authentication credentials. Such an attack would also enable the rogue BS to impersonate name servers, allowing it to redirect user requests to computers with malware without easy detection. This vulnerability is mitigated in IEEE 802.16e-2005 and IEEE 802.16-2009 by the use of mutual authentication.

- **Weak encryption algorithms.** For encrypting communications, IEEE 802.16-2004 only supports the use of DES-CBC, which has well-documented weaknesses and is no longer approved for Federal agency use in protecting communications. IEEE 802.16e-2005 and IEEE 802.16-2009 support DES-CBC, but they also support multiple modes of AES that are approved for Federal government use.

- **Interjection of reused TEKs.** IEEE 802.16-2004 TEKs employ a 2-bit encryption sequence identifier to determine which TEK is actively used to secure communications. A 2-bit identifier permits only four possible identifier values, rendering the system vulnerable to replay attacks. The interjection of reused TEKs may lead to the disclosure of data and the TEK to unauthorized parties [Joh04]. This concern is resolved in IEEE 802.16e-2005 and IEEE 802.16-2009 with the introduction of AES-CCM, which provides per packet randomization by adding a unique packet number to each data packet to protect the integrity of data and portions of the packet header.

- **Unencrypted management messages.** Management messages are not encrypted and are susceptible to eavesdropping attacks. Encryption is not applied to these messages to increase the efficiency of network operations. IEEE 802.16-2004 does not provide any data authenticity protection for management messages. IEEE 802.16e-2005 and IEEE 802.16-2009 provide integrity protection for certain unicast management messages by appending a unique digest to protect against malicious replay or modification attacks. This digest is not added to IEEE 802.16 multicast and initial network entry management messages. As with all wireless systems, digest integrity protection cannot be applied to management messages sent to multiple recipients (i.e., multicast transmissions), and initial network entry management messages cannot leverage integrity protection because nodes must first be authenticated to create the unique digest. In a man-in-the-middle attack (see Section 4.2), an

---

[30] These controls are typically documented in a System Security Plan. For more information, see NIST SP 800-18 Revision 1, *Guide for Developing Security Plans for Federal Information Systems* (http://csrc.nist.gov/publications/nistpubs/800-18-Rev1/sp800-18-Rev1-final.pdf).

adversary may manipulate management messages to disrupt network communications, by denial of service (DoS) attacks aimed at the WiMAX system, at specific network nodes, or both. IEEE 802.16m supports full encryption and authentication for management messages, but it has not been integrated into many vendor solutions as of the writing of this publication because IEEE 802.16m was recently released.

- **Use of electromagnetic spectrum as a communications medium.** Using RF to communicate inherently enables execution of a DoS attack by introducing a powerful RF source intended to overwhelm system radio spectrum. This vulnerability is associated with all wireless technologies. The only defenses are either to locate and remove the source of RF interference or to move to another channel. Such actions can be challenging because of the large coverage areas of WMANs and the scarcity of alternative frequencies to support communications. It is recommended that organizations plan for out-of-band communications in the event of a DoS attack.

4.2 Threats

WiMAX network threats focus on compromising the radio links between WiMAX nodes. LOS WiMAX systems pose a greater challenge to attack compared with NLOS systems because an adversary would have to physically locate equipment between the transmitting nodes to compromise the confidentiality or integrity of the wireless link. NLOS systems provide wireless coverage over large geographic regions, which expands the potential staging areas for both clients and adversaries. The following threats affect all WiMAX systems:

- **RF jamming.** All wireless technologies are susceptible to RF jamming attacks. The threat arises from an adversary introducing a powerful RF signal to overwhelm the spectrum being used by the system, thus denying service to all wireless nodes within range of the interference. RF jamming is classified as a DoS attack. The risk associated with this threat is identical for IEEE 802.16-2004, IEEE 802.16e-2005, and IEEE 802.16-2009 WiMAX systems.

- **Scrambling.** Scrambling attacks, which are the precise injections of RF interference during the transmission of specific management messages, affect all wireless systems. These attacks prevent proper network ranging and bandwidth allocations with the intent to degrade overall system performance [Nas08]. Scrambling attacks are more difficult to identify than jamming attacks because they are engaged for short time periods and are not a constant source of interference. The risk associated with this threat is identical for IEEE 802.16-2004, IEEE 802.16e-2005, and IEEE 802.16-2009.

- **Subtle management message manipulation.** Exploitation of unauthenticated management messages can result in subtle DoS, replay, or misappropriation[31] attacks that are difficult to detect. These attacks spoof management messages to make them appear as though they come from a legitimate BS or SS/MS, allowing them to deny service to various nodes in the WiMAX system. A *water torture* attack is an example of a subtle DoS in which an adversary drains a client node's battery by sending a constant series of management messages to the SS/MS [Joh04]. IEEE 802.16e-2005 and IEEE 802.16-2009 provide integrity protection for certain unicast management messages following initial network registration with an appended integrity protection digest. All other IEEE 802.16e-2005 and IEEE 802.16-2009 management messages, and all IEEE 802.16-2004 management messages, are susceptible to attacks involving manipulation.

- **Man-in-the-middle.** Man-in-the-middle attacks occur when an adversary deceives an SS/MS to appear as a legitimate BS while simultaneously deceiving a BS to appear as a legitimate SS/MS. This

---

[31] Misappropriation occurs when an attacker steals or makes unauthorized use of a service.

may allow an adversary to act as a pass-through for all SS/MS communications and to inject malicious traffic into the communications stream. An adversary can perform a man-in-the-middle attack by exploiting unprotected management messages during the initial network entry process. This is because the management messages that negotiate an SS's/MS's security capabilities are not protected. If an adversary is able to impersonate a legitimate party to both the SS/MS and BS, an adversary could send malicious management messages and negotiate weaker security protection between the SS/MS and BS [Han06]. This weaker security protection may allow an adversary to eavesdrop and corrupt data communications. Mandating the use of AES-CCM in IEEE 802.16e-2005 and IEEE 802.16-2009 helps mitigate this attack because it appends a unique value to each data packet, which, in turn, prevents the man-in-the-middle traffic relays between BS and SS/MS. IEEE 802.16-2004 does not offer adequate protection against man-in-the-middle attacks.

■ **Eavesdropping.** Eavesdropping occurs when an adversary uses a WiMAX traffic analyzer within the range of a BS or SS/MS. The large operating range of WiMAX networks helps to shield eavesdroppers from detection; eavesdropping mitigation relies heavily on technical controls that protect the confidentiality and integrity of communications. The adversary may monitor management message traffic to identify encryption ciphers, determine the footprint of the network, or conduct traffic analysis regarding specific WiMAX nodes. Data messages collected during eavesdropping can also be used to decipher DES-CBC encryption; however, AES provides robust data message confidentiality that cannot be circumvented through eavesdropping. The risk associated with eavesdropping management messages is identical for IEEE 802.16-2004, IEEE 802.16e-2005, and IEEE 802.16-2009. The risk associated with eavesdropping data messages is significant for IEEE 802.16-2004 systems due to weak encryption. IEEE 802.16e-2005 and IEEE 802.16-2009 systems offer the stronger AES cipher to protect data messages from eavesdropping.

4.3 Countermeasures

This section presents countermeasures that may be used to reduce or mitigate the risks inherent to WiMAX systems. These countermeasures do not guarantee security and cannot prevent all possible attacks. The optimum security design is a dynamic intersection of threat risk and the cost of countermeasures that will change in response to technology. Organizations should implement countermeasures commensurate with their acceptable level of risk.

The WiMAX management, operational, and technical countermeasures described in the following sections take an approach similar to that of NIST SP 800-48 Revision 1, *Guide to Securing Legacy IEEE 802.11 Wireless Networks* [Sca08]. IEEE 802.11 and IEEE 802.16 share many management and operational controls but strongly differ in their technical controls. WiMAX systems should leverage the countermeasures found in this Special Publication, in FIPS PUB 199, and in NIST SP 800-53. FIPS PUB 199, *Standards for Security Categorization of Federal Information and Information Systems* establishes three security categories—low, moderate, and high—based on the potential impact of a security breach involving a particular system [NIS04]. NIST SP 800-53, *Recommended Security Controls for Federal Information Systems and Organizations*, provides recommendations for minimum management, operational, and technical security controls for information systems based on the FIPS PUB 199 impact categories [NIS09]. Regardless of a system's sensitivity classification, WiMAX security should be incorporated throughout the entire lifecycle of WiMAX solutions [Kis08].

4.3.1 Management Countermeasures

Management countermeasures generally address any problem related to risk, system planning, or security assessment by an organization's management. Organizations should develop a wireless security policy that addresses WiMAX technology. A security policy is an organization's foundation for designing, implementing, and maintaining properly secured technologies. WiMAX policy should address the design and operation of the technical infrastructure and the behavior of users. Policy considerations for WiMAX systems should include the following:

- Roles and responsibilities

    - Which users or groups of users are authorized to use the WiMAX system

    - Which office or officer provides the strategic oversight and planning for all WiMAX technology programs

    - Which parties are authorized and responsible for installing and configuring WiMAX equipment

    - Which individual or entity tracks the progress of WiMAX security standards, features, threats, and vulnerabilities to help ensure continued secure implementation of WiMAX technology

    - Which individual or entity is responsible for incorporating WiMAX technology risk into the organization's risk management framework[32]

- WiMAX infrastructure

    - Physical security requirements for WiMAX assets

    - The use of standards-based WiMAX system technologies

    - Types of information permitted over the WiMAX system, including acceptable use guidelines

    - How WiMAX transmissions should be protected, including requirements for the use of encryption and for cryptographic key management

    - A mitigation plan or transition plan for legacy or WiMAX systems that are not compliant with Federal security standards

    - Inventory of IEEE 802.16 BSs, SSs/MSs, and other devices

- WiMAX client device security

    - Conditions under which WiMAX client devices are allowed to be used and operated

    - Standard hardware and software configurations that must be implemented on WiMAX devices to ensure the appropriate level of security

    - Standard operating procedures (SOP) for reporting lost or stolen WiMAX client devices

- WiMAX security assessments[33]

    - Frequency and scope of WiMAX security assessments

    - Standardized approach to vulnerability assessment, risk statements, risk levels, and corrective actions

---

[32] For more information and guidance on organizational information system risk management, see NIST SP 800-37 Revision 1, *Guide for Applying the Risk Management Framework to Federal Information Systems A Security Life Cycle Approach* (http://csrc.nist.gov/publications/PubsSPs.html).

[33] Security assessments, or system audits, are essential tools for checking the security posture of an IEEE 802.16 WMAN and for determining the corrective actions to make sure such systems remain secure. It is important for organizations to perform regular system audits using wireless and vulnerability assessment tools. For more information on network security, see NIST SP 800-115, *Technical Guide to Information Security Testing and Assessment* (http://csrc.nist.gov/publications/PubsSPs.html).

4.3.2 Operational Countermeasures

Operational countermeasures include controls that are executed by people, e.g., personnel security, physical environment protection, configuration management, security awareness and training, and incident response. These controls are documented in a system security plan (SSP) which should be maintained by all parties involved with WiMAX system operations. SSPs are living documents that provide an overview of the security requirements of a system and describe the controls in place to meet those requirements; this includes all system hardware and software, policies, roles and responsibilities, and other documentation materials. Documentation itself is a security control, as it formalizes security and operational procedures to a given system.[34]

Physical security is fundamental to ensuring that only authorized personnel have access to WiMAX equipment. Physical security includes measures such as physical access control systems, personnel security and identification, and external boundary protection. For example, integrating Federal personal identity verification (PIV)[35] into physical access controls can reduce the risk of unauthorized access to WiMAX systems.[36] WiMAX system administrators and users should receive training to address the specific challenges and threats to wireless technologies. While it is difficult to prevent unauthorized users from attempting to access a WiMAX system because of its expansive coverage area, the use of additional security mechanisms may help prevent the theft, alteration, or misuse of WiMAX infrastructure components.

WiMAX operates on licensed or unlicensed RF spectrum. The most prevalent spectrum used to accommodate WiMAX is the 2.5 GHz licensed range, but WiMAX solutions are also viable across several unlicensed spectrum ranges. Organizations should understand the implications of spectrum allocation as it impacts system availability. Due to the proliferation of unlicensed wireless technologies, interference may become an implementation obstacle when operating in unlicensed spectrum. Regardless of which spectrum frequency is used, organizations should use counter-interference technologies[37] in addition to site surveys to ensure system availability.

Prior to deployment, site surveys construct the foundation for a WiMAX system's design to ensure system availability. Long distance radio transmissions should be tailored and optimized for RF obstacles and interference sources. Site surveys help limit range to provide an organization operational awareness of a system's coverage area. Site survey tools include terrain maps, global positioning systems, RF propagation models, spectrum analyzers, packet analyzers, and additional tools which provide a more thorough understanding of the environment's RF landscape. Conducting a WMAN site survey requires specialized skills, and it is typically provided as part of the overall vendor solution. Organizations should, at a minimum, involve themselves in the site survey process and document its findings in the system security plan.

As with all wireless technologies, operational countermeasures may not provide protection against general wireless threats such as DoS, eavesdropping, man-in-the-middle, and message replay. Operational controls often require highly specialized expertise and rely upon both management and technical controls.

---

[34] For more information and guidance on system documentation controls, see NIST SP 800-18, Revision 1, *Guide for Developing Security Plans for Federal Information Systems* (http://csrc.nist.gov/publications/PubsSPs.html).

[35] FIPS PUB 201 specifies the architecture and technical requirements for a common identification standard for Federal employees and contractors. For more information see http://csrc.nist.gov/publications/fips/fips201-1/FIPS-201-1-chng1.pdf.

[36] For more information and guidance on PIV integration with physical access controls, see NIST SP 800-116, A *Recommendation for the Use of PIV Credentials in Physical Access Control Systems (PACS)* (http://csrc.nist.gov/publications/PubsSPs.html).

[37] Examples of counter-interference technologies include, but are not limited to, Dynamic Frequency Selection (DFS); Multiple Input, Multiple Output (MIMO); and Adaptable Antenna Support (AAS).

4.3.3 Technical Countermeasures

Technical countermeasures are system safeguards implemented by computer systems, such as, authentication, access control, auditing, and communication protection. Technical countermeasures are typically designed into WiMAX systems before implementation and vary widely between vendors. Before implementing a WiMAX system, an organization should consult WiMAX vendors to gain a better understanding of potential system reconfiguration constraints and the need for compensating controls to address technical security needs that the WiMAX product may not address.

Confidentiality and Integrity Protection

WiMAX systems broadcast data over a large geographic area that is usually outside the organization's physical control. Organizations must rely on data-in-transit encryption to provide confidentiality and integrity protection for its wireless links. As explained in Section 3.4, each IEEE 802.16 standard specifies encryption algorithms to provide data protection.

Federal agencies that need to protect the confidentiality of their WiMAX data communications are required to use FIPS-approved cryptographic modules implemented by encryption algorithms to protect data communications. This can be accomplished in one of two ways: using WiMAX equipment (BSs, SSs, and MSs) that is FIPS-validated and operating in FIPS mode,[38] or encrypting the data through a separate encryption technology, such as a FIPS-validated VPN. Organizations using IEEE 802.16e-2005 or IEEE 802.16-2009 WiMAX technology should implement FIPS-validated AES encryption to protect OSI Data Link Layer radio data communications. Organizations using WiMAX solutions that do not support FIPS-validated encryption schemes (e.g., IEEE 802.16-2004 technologies, IEEE 802.16e-2005 and IEEE 802.16-2009 technologies that are not FIPS-validated) should implement FIPS-validated encryption overlay solutions to protect OSI Data Link or Network Layer data communications.[39] Data Link encryption overlay solutions provide encryption for all data traffic taking place at the OSI Network Layer and above, including packet headers and trailers. Network Layer encryption solutions provide encryption for the data portion of the Network Layer, including all upper-layer data.

Encryption overlay solutions require an encryption appliance behind the BS and a corresponding supplicant or termination point on the end user device. A supplicant is a software program on a client device that coordinates authentication with the encryption overlay appliance. The encryption appliance and supplicant work in tandem to encrypt data before it reaches the WiMAX system. Thus, the WiMAX system serves as the transmission medium between the supplicant and the appliance. Figure 4-1 illustrates this communications path over the WiMAX network.

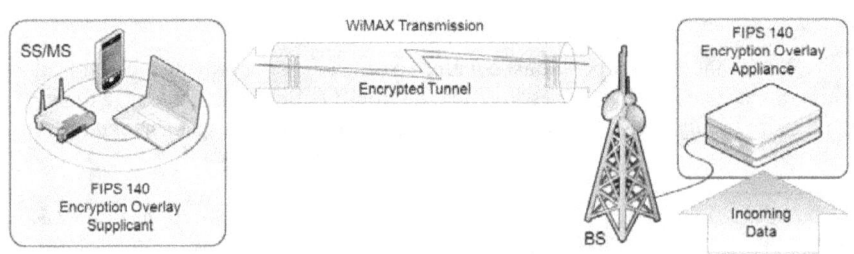

Figure 4-1. Encryption Overlay Solution

---

[38] FIPS 140-2 is the current Federal standard that specifies the security requirements that will be satisfied by a cryptographic module. A new version, FIPS 140-3, is in development. Both are at http://csrc.nist.gov/publications/PubsFIPS.html.

[39] The use of encryption overlay solutions may limit mobile operations because of a lack of handover and QoS support.

Data Link Layer encryption solutions are not guaranteed to work with all WiMAX solutions because of the way IEEE 802.16 solutions treat the OSI Data Link packet convergence sub-layer that is responsible for transferring data between the OSI Data Link and Network layers. IEEE 802.16-2004, IEEE 802.16e-2005, and IEEE 802.16-2009 allow for two traffic types, Ethernet header and IP. Solutions that use IP type may not work with solutions that completely encrypt the OSI Network Layer because the layered encryption will hide information required to pass packets through the WiMAX system. Typically, IP classification is used in mobile operations while Ethernet header classification is used in P2P and PMP operations. Prior to procuring an OSI Data Link Layer encryption solution, organizations should first inquire which traffic classification a vendor's WiMAX implementation supports.

Another method of providing confidentiality and integrity protection is by using a VPN. A VPN is a virtual network that creates a secure tunnel between devices to provide a secure communications channel for data and IP information. VPNs are often used to facilitate the secure transfer of sensitive data across shared and untrusted networks. VPNs can be established within a WMAN to create a private network and protect sensitive communications from other parties sharing network resources. VPNs are a mature technology, and a variety of VPN technologies exist, such as IPsec and Secure Sockets Layer (SSL) VPNs.[40]

Organizations using VPNs to protect the confidentiality of WMAN communications should configure the VPNs to use FIPS-validated encryption algorithms employing FIPS-validated cryptographic modules.[41] One method is to use VPNs to protect WMAN communications by establishing a VPN tunnel between the WMAN client device (i.e., the SS/MS) and a VPN concentrator behind the BS. With an IPsec VPN, security services are provided at the OSI Network Layer of the protocol stack, which will secure all applications and protocols operating at the OSI Network Layer and above. VPN security services are independent of OSI Data Link Layer protocols, and they are recommended for use if the underlying wireless security mechanisms are vulnerable or lack FIPS validation. VPNs do not eliminate all risk from wireless networking. If further assurance is required, OSI Data Link wireless security controls can be used to provide an additional layer of security.

When determining how confidentiality and integrity should be protected, organizations should examine their WiMAX client devices to determine which types of cryptographic algorithms are supported and their key length limitations and re-keying period. Organizations should verify that these client cryptographic capabilities meet the minimum confidentiality protection requirements. Secret keys should be periodically replaced in accordance with an organization's IT security policy to reduce the potential effect of a key compromise.[42]

Authentication and Authorization

IEEE 802.16e-2005 and IEEE 802.16-2009 support an array of authentication solutions to provide both device and user authentication between a BS and MS/SSs. These authentication solutions can support the use of usernames and passwords, smart cards, subscriber identity module cards, biometrics, public key infrastructure (PKI), or a combination of these solutions (e.g., smart cards with PKI).[43] Organizations should strongly consider WiMAX solutions capable of supporting Extensible Authentication Protocol (EAP) methods for mutual authentication as discussed in NIST SP 800-120, *Recommendation for EAP*

---

[40] For more information and guidance on VPNs, see NIST SP 800-77, *Guide to IPSec VPNs* and NIST SP 800-113, *Guide to SSL VPNs* (http://csrc.nist.gov/publications/PubsSPs.html).

[41] Use of such modules and algorithms is mandatory for Federal agencies in this situation.

[42] For more information on key management, see NIST SP 800-57, *Recommendation for Key Management* (http://csrc.nist.gov/publications/PubsSPs.html).

[43] For more information on entity authentication using public key cryptography, see FIPS PUB 196, *Entity Authentication Using Public Key Cryptography* (http://csrc.nist.gov/publications/fips/fips196/fips196.pdf).

Methods Used in Wireless Network Access Authentication.[44] WiMAX solutions that cannot meet these criteria should employ a different means of authentication at a higher layer (e.g., encryption overlay or VPN).

Client Device Security

Devices that have been granted access to a WiMAX system should be properly secured to enhance the system security posture. Securing the infrastructure without properly securing client devices may compromise system security. Organizations should implement the client security controls appropriate for the type of device, the device's operating system and applications, the sensitivity of the data that the device holds and accesses, and the threats that it faces. Possible client device security considerations include the following:

- **Personal firewalls.** Resources on WMANs have a higher risk of attack than those on wired networks, because they generally do not have the same degree of protection. Personal firewalls increase device security by offering some protection against unwanted network connections initiated by other hosts. Personal firewalls are software applications that reside on a client device and are either client managed or centrally managed. Centrally managed solutions provide a greater degree of protection because IT departments may configure and remotely manage these solutions rather than leaving the management to the end user. Centrally managed solutions allow organizations to modify client firewalls to protect against known vulnerabilities, react to new threats and vulnerabilities, and maintain a consistent security policy for all client users. Personal firewalls may also have built-in VPN capabilities that further ease deployment.

- **Host-based intrusion detection and prevention systems (IDPS).** A host-based IDPS provides complementary security services to a personal firewall. Host-based IDPS software monitors and analyzes the internal state of a client device. IDPS products review logs to ensure that the system and applications are not functioning unexpectedly, such as applications inexplicably accessing or altering other portions of the system. Several host-based IDPS software products also monitor inbound and outbound network communications and report or possibly block suspicious activity.

- **Antimalware software.** Antivirus software and other forms of antimalware software can assist in preventing the spread of viruses, worms, and other malware between networked devices. Most client devices are at risk from malware threats, so such devices should have appropriate antimalware software installed and receive automatic updates.

- **IEEE 802.16 radio management.** Client devices that have no business need for IEEE 802.16 should have their wireless radios disabled by default. If IEEE 802.16 use is required, users should disable the radio when not in use (if feasible).

- **Policy enforcement.** Client devices should be configured to comply with implemented WMAN policies. Devices should be configured according to policy, such as disabling services or altering default configurations. In addition, policy-driven software solutions can be implemented on client devices to prevent or allow certain actions when specific parameters exist. Policy-driven software helps ensure that client devices and users comply with an organization's defined policies. For example, policy-based software can prevent client devices from having more than one network interface enabled at a time.

In addition to these considerations for client device security, organizations should also ensure that wireless client devices are logically separated from the organization's wired networks. This is most

---

[44]   http://csrc.nist.gov/publications/PubsSPs.html

commonly implemented by using a filtering device between the WiMAX system and the wired network. A network appliance can logically segment user populations and govern data communications.

Patches, Upgrades, and Updates

Like other network equipment vendors, IEEE 802.16 vendors typically issue patches, upgrades, or firmware updates to correct known software and hardware security vulnerabilities. Network administrators should regularly check with vendors to identify these updates and implement them as needed, in accordance with the organization's procedures and policies.[45] In addition, many vendors have security alert email lists to advise customers of new vulnerabilities and attacks. Administrators can also check the National Vulnerability Database[46] and Wireless Vulnerability & Exploits database[47] for listings of publicly known vulnerabilities.

4.4 Vulnerabilities, Threats, and Countermeasures Summary Table

IEEE 802.16e-2005 and IEEE 802.16-2009 include security features that address many of the technical shortcomings of IEEE 802.16-2004. Organizations with WiMAX implementations pre-dating IEEE 802.16e-2005 should implement a mitigation plan and/or migration plan to outline the procedures or restrictions for transitioning legacy WiMAX systems to a compliant security architecture.

Table 4-1 presents a summary of the IEEE 802.16-2004, IEEE 802.16e-2005, and IEEE 802.16-2009 vulnerabilities, threats, and countermeasures discussed throughout this section.

Table 4-1. Vulnerabilities, Threats, and Countermeasures Summary

Security Concern or Vulnerability	Threat Discussion	Countermeasure
IEEE 802.16-2004 Based WiMAX Systems		
Unilateral authentication of SS by BS	SSs have no method for verifying the identity of BSs. This leaves SSs susceptible to forgery attacks by a rogue BS. This may result in dogradod porformanco, information theft, or DoS attacks. In addition, this authentication schema leaves a system susceptible to man-in-the-middle attacks.	Force communications to take place over a VPN or encryption overlay that authenticates the devices/users outside of the WiMAX system's native controls.
DES-CBC Weakness	DES-CBC is a weak algorithm that cannot ensure confidentiality of data. Using DES-CBC may lead to unauthorized disclosure of information. Threats may include DoS, eavesdropping, and man-in-the-middle attacks.	Implement an encryption overlay or VPN that employs a FIPS-validated solution, i.e., FIPS-validated encryption algorithms and FIPS-validated cryptographic modules

---

[45] For more information on patch and vulnerability management, see NIST SP 800-40 Version 2, *Creating a Patch and Vulnerability Management Program* (http://csrc.nist.gov/publications/nistpubs/800-40-Ver2/SP800-40v2.pdf).

[46] The NIST National Vulnerability Database can be found at http://nvd.nist.gov/.

[47] The Wireless Vulnerability & Exploits database can be found at http://www.wve.org/.

Security Concern or Vulnerability	Threat Discussion	Countermeasure
Interjection of Reused TEK	The short two-bit TEK identifier wraps to zero on every fourth re-key. This allows adversaries to reuse expired TEKs and perform replay attacks leading to unauthorized disclosure of information and compromise of the TEK.	Implement encryption overlay or VPN that employs a FIPS-validated solution, i.e., FIPS-validated encryption algorithms and FIPS-validated cryptographic modules
All WiMAX Systems		
Unencrypted Management Messages	BSs and SSs/MSs communicate using unencrypted management messages to facilitate network entry, node registration, bandwidth allocation, and ranging. These messages are not encrypted. Integrity checks are added to unicast messages to prevent replay attacks; however, non-unicast messages are still vulnerable to DoS attacks. Unencrypted management messages are subject to eavesdropping, replay attacks, scrambling, and subtle manipulation aimed at degrading service. If a system is not using AES-CCM, it remains vulnerable to a man-in-the-middle attack.	There is no threat mitigation for unencrypted management messages; however, AES-CCM helps to mitigate against man-in-the-middle attacks. The impact from management message exploitation is exposure of node registration information and various DoS attacks. It is recommended that organizations plan for out-of-band communications in the event of a DoS attack. At a minimum, organizational SOPs should include DoS incident response plans.
Lack of Native FIPS-Validated Solutions	As of mid-2010, most WiMAX vendors had not completed the FIPS-validation process. This prevents Federal entities from relying on native WiMAX security until vendors complete the validation process. Communications relying solely on unvalidated native WiMAX security to protect confidentiality may be vulnerable to encryption implementation weaknesses that may lead to unauthorized disclosure of information. These threats include eavesdropping, man-in-the-middle, and DoS attacks.	Implement encryption overlay or VPN that employs a FIPS-validated solution, i.e., FIPS-validated encryption algorithms and FIPS-validated cryptographic modules.
Use of Wireless as a Communications Medium	DoS attacks can be executed by the introduction of a powerful RF source intended to overwhelm system radio spectrum.	Locate and remove the source of RF interference. This can be challenging because of the large coverage areas of WMANs. It is recommended that organizations plan for out-of-band communications in the event of a DoS attack.

Appendix A—Glossary

Selected terms used in the guide are defined below.

Authentication: For the purposes of this guide, the process of verifying the identity claimed by a WiMAX *device*. *User* authentication is also an option supported by IEEE 802.16e-2005.

Authorization: The process that takes place after authentication is complete to determine which resources/services are available to a WiMAX device.

Authorization key (AK): A key exchanged between the base station and subscriber station/mobile station to authenticate one another prior to the traffic encryption key (TEK) exchange.

Authorized data security association (SA) list: A list that the BS provides to the SS/MS that indicates which data encryption SAs the SS/MS is authorized to use.

Backhaul: Typically a high capacity line from a remote site or network to a central site or network.

Base station (BS): The node that logically connects fixed and mobile subscriber stations to operator networks. The BS governs access to the operator networks and maintains communications with client devices. A BS consists of the infrastructure elements necessary to enable wireless communications, i.e., antennas, transceivers, and other electromagnetic wave transmitting equipment. BSs are typically fixed nodes, but in a tactical environment, they may also be considered mobile.

Confidentiality: For the purposes of this guide, prevention of the disclosure of information by ensuring that only authorized devices can view the contents of WiMAX communications.

Data encryption security association (SA) type indicator: An indicator defining the type of data encryption SA (primary, static, or dynamic).

Diameter: A successor AAA protocol to RADIUS that supports enhanced security and communication methods.

Eavesdropping: Type of attack in which an adversary uses a WiMAX traffic analyzer within the range of a BS or SS/MS to monitor WiMAX communications.

Group key encryption key (GKEK): A cryptographic key used to encrypt the GTEK sent in multicast messages between a BS and two or more SSs/MSs.

Group traffic encryption key (GTEK): A cryptographic key used to encrypt multicast traffic between a BS and two or more SSs/MSs.

Internet Protocol Security (IPsec): An OSI Network layer security protocol that provides authentication and encryption over IP networks.

Intrusion detection and prevention system (IDPS): An appliance or software product that provides complementary security services to a personal firewall, monitoring and analyzing the internal state of a client device. IDPS products review logs to ensure that the system and applications are not functioning unexpectedly, such as applications inexplicably accessing or altering other portions of the system. Several host-based IDPS software products also monitor inbound and outbound network communications and report or possibly block suspicious activity.

Jitter: As it relates to queuing, the difference in latency of packets.

Key encryption key (KEK): A key derived from the authorization key that is used to encrypt traffic encryption keys (TEK) during the TEK exchange.

Last mile broadband access: Communications technology that bridges the transmission distance between the broadband service provider infrastructure and the customer premises equipment.

Line-of-sight (LOS) signal propagation: Electromagnetic signaling that is highly sensitive to radio frequency obstacles and therefore requires an unobstructed view between transmitting stations.

Management countermeasure: A countermeasure that addresses any concern related to risk, system planning, or security assessment by an organization's management.

Management message: A message used for maintaining communications between an SS/MS and BS, i.e., establishing communication parameters, exchanging privacy settings, and performing system registration events (initial network entry, handoffs, etc.). These messages are not encrypted and are susceptible to eavesdropping attacks.

Man-in-the-middle (MITM): An attack that occurs when an adversary deceives an SS/MS to appear as a legitimate BS while simultaneously deceiving a BS to appear as a legitimate SS/MS. This may allow an adversary to act as a pass-through for all communications and to inject malicious traffic into the communications stream.

Message authentication key: A key that validates the data authenticity of the key distribution messages sent from the BS to the SS/MS.

Misappropriation: An attack in which the attacker steals or makes unauthorized use of a service.

Mobile subscriber (MS): As defined in IEEE 802.16e-2005, an SS capable of moving at vehicular speeds and that supports enhanced power management modes of operation. MS devices typically have a small form factor and are self-powered, e.g., laptops, ultra-mobile portable computers, cellular telephones, or other portable electronic devices.

Mobile topology: A configuration similar to a cellular network, where multiple BSs collaborate and provide seamless communications over a distributed network to both SSs and MSs.

Multi-hop relay topology: A configuration that extends a BS's coverage area by permitting SSs and MSs to relay traffic by acting as RSs. Data destined to an SS/MS outside of the BS's range is relayed through adjacent RSs.

Multiple input, multiple output (MIMO) technology: The use of multiple antennas and advanced signaling techniques to increase wireless network range, resiliency, and speed.

Non-line-of-sight (NLOS) signal propagation: Electromagnetic signaling that uses advanced modulation techniques to compensate for signal obstacles and allows indirect communications between transmitting stations.

Operational countermeasure: A countermeasure that includes controls that are executed by people, e.g., physical environment protection, configuration management, and incident response.

Personal firewall: A software application residing on a client device that increases device security by offering some protection against unwanted network connections initiated by other hosts. Personal firewalls may be client managed or centrally managed.

Quality of Service (QoS): A categorization of different types of network traffic to prioritize latency-sensitive data over non-latency-sensitive data.

Radio frequency (RF) jamming: A threat in which an adversary introduces a powerful RF signal to overwhelm the spectrum being used by the system, thus denying service to all wireless nodes within range of the interference. RF jamming is classified as a DoS attack.

Relay station (RS): An SS that is configured to forward traffic to other stations in a multi-hop Security Zone.

Remote Authentication Dial-In User Service (RADIUS): A centralized Authentication, Authorization, and Accounting (AAA) protocol currently defined in RFC 2865.

Scrambling: The precise injection of RF interference during the transmission of specific management messages. These attacks prevent proper network ranging and bandwidth allocations with the intent to degrade overall system performance.

Security association (SA): The logical set of security parameters containing elements required for authentication, key establishment, and data encryption.

Security association identifier (SAID): A unique 16-bit value that identifies the SA.

Security Zone (SZ): A set of trusted relationships between a BS and a group of RSs.

Standard operating procedure: A set of instructions used to describe a process or procedure that performs an explicit operation or explicit reaction to a given event.

Subscriber station (SS): A wireless node that typically communicates only with a BS, except when part of a multi-hop relay configuration.

System security plan (SSP): A system document that provides an overview of the security requirements of a system and describes the controls in place to meet those requirements.

Technical countermeasures: System safeguards implemented by computer systems, including controls such as authentication, access control, auditing, and protecting communications.

Unilateral authentication: An IEEE 802.16-2004 vulnerability resulting from PKMv1 providing for authentication of SSs by BSs but not for authentication of BSs by SSs. Lack of mutual authentication may allow a rogue BS to impersonate a legitimate BS, thereby rendering the SS unable to verify the authenticity of protocol messages received from the BS. This may enable a rogue BS operator to degrade performance or steal valuable information by conducting DoS or man-in-the-middle attacks against client SSs.

Virtual private network (VPN): A logical network that is established at the network layer of the OSI model. The logical network typically provides authentication and data confidentiality services for some subset of a larger physical network.

WiMAX: A wireless metropolitan area network (WMAN) technology based on the IEEE 802.16 family of standards used for a variety of purposes, including, but not limited to, fixed last-mile broadband access, long-range wireless backhaul, and access layer technology for mobile wireless subscribers operating on telecommunications networks.

Appendix B—Acronyms and Abbreviations

Selected acronyms and abbreviations used in the guide are defined below.

AAA	Authentication, Authorization, and Accounting
AAS	Adaptable Antenna Support
AES	Advanced Encryption Standard
AK	Authorization Key
AKID	Authorization Key Identifier
BS	Base Station
CBC	Cipher Block Chaining
CCM	Counter with CBC Message Authentication Code
CFB	Cipher Feedback
CTR	Counter
DES	Digital Encryption Standard
DFS	Dynamic Frequency Selection
DoS	Denial of Service
EAP	Extensible Authentication Protocol
ECB	Electronic Codebook
EIK	EAP Integrity Key
FCC	Federal Communications Commission
FIPS	Federal Information Processing Standard
FISMA	Federal Information Security Management Act
Gbps	Gigabits per second
GHz	Gigahertz
GKEK	Group Key Encryption Key
GTEK	Group Traffic Encryption Key
IDPS	Intrusion Detection and Prevention System
IETF	Internet Engineering Task Force
IPsec	Internet Protocol Security
ISO	International Organization for Standardization
IT	Information Technology
ITL	Information Technology Laboratory
IV	Initialization Vector
KEK	Key Encryption Key
LAN	Local Area Network
LOS	Line of Sight
MAC	Media Access Control
MAN	Metropolitan Area Network
Mbps	Megabits per second
MIB	Management Information Base

MIMO	Multiple Input, Multiple Output
MS	Mobile Subscriber
MSK	Master Session Key
NIST	National Institute of Standards and Technology
NLOS	Non-Line-of-Sight
OFB	Output Feedback
OMB	Office of Management and Budget
OSI	Open Systems Interconnection
P2P	Point-to-Point
PIV	Personal Identity Verification
PKI	Public Key Infrastructure
PKM	Privacy Key Management
PKMv1	Privacy Key Management Protocol version 1
PKMv2	Privacy Key Management Protocol version 2
PMK	Pairwise Master Key
PMP	Point-to-Multipoint
PN	Packet Number
Pre-PAK	Pre-Primary Authorization Key
PUB	Publication
QoS	Quality of Service
RADIUS	Remote Authentication Dial-in User Service
RF	Radio Frequency
RS	Relay Station
RSA	Rivest, Shamir, and Adleman
SA	Security Association
SAID	Security Association Identifier
SOP	Standard Operating Procedures
SS	Subscriber Station
SSL	Secure Sockets Layer
SSP	System Security Plan
SZ	Security Zone
TEK	Traffic Encryption Key
VoIP	Voice over Internet Protocol
VPN	Virtual Private Network
WLAN	Wireless Local Area Network
WMAN	Wireless Metropolitan Area Network

Appendix C—References

The list below provides references for this publication.

[Fra07]   S. Frankel et al, NIST Special Publication 800-97, Establishing Wireless Robust Security
          Networks: A Guide to IEEE 802.11i, NIST, 2007.
          http://csrc.nist.gov/publications/nistpubs/800-97/SP800-97.pdf

[Han06]   Tao Han et al, Analysis of Mobile WiMAX Security: Vulnerabilities and Solutions, Key Lab of
          Universal Wireless Communications, Ministry of Education, Beijing University of Posts and
          Telecommunications, 2007.

[IEE04]   IEEE Standard 802.16-2004, IEEE Standard for Local and Metropolitan Area Networks—Part
          16: Air Interface for Fixed Broadband Wireless Access Systems, IEEE, 2004.

[IEE05]   IEEE Standard 802.16e-2005, IEEE Standard for Local and Metropolitan Area Networks—Part
          16: Air Interface for Fixed and Mobile Broadband Wireless Access Systems Amendment 2:
          Physical and Medium Access Control Layers for Combined Fixed and Mobile Operation in
          Licensed Bands, IEEE, 2005.

[IEE09]   IEEE Standard 802.16-2009, IEEE Standard for Local and Metropolitan Area Networks—Part
          16: Air Interface for Broadband Wireless Access Systems, IEEE, 2009.

[IEJ09]   IEEE Standard 802.16j-2009, IEEE Standard for Local and Metropolitan Area Networks—Part
          16: Air Interface for Broadband Wireless Access Systems: Amendment 1: Multiple Relay
          Specification, IEEE, 2009.

[Joh04]   D. Johnston and J. Walker, Overview of IEEE 802.16 Security, *IEEE Security and Privacy*,
          Vol. 2, no. 3, pp 40-48, May/June 2004.

[Kis08]   R. Kissel et al, NIST Special Publication 800-64 Revision 2, Security Considerations in the
          System Development Life Cycle, NIST, 2008. http://csrc.nist.gov/publications/nistpubs/800-
          64-Rev2/SP800-64-Revision2.pdf

[Nas08]   Mahmoud Nasreldin et al, WiMAX Security, 22nd International Conference on Advanced
          Information Networking and Applications, 2008.

[NIS04]   FIPS PUB 199, Standards for Security Categorization of Federal Information and Information
          Systems, NIST, 2004. http://csrc.nist.gov/publications/fips/fips199/FIPS-PUB-199-final.pdf

[NIS09]   NIST Special Publication 800-53 Revision 3, Recommended Security Controls for Federal
          Information Systems and Organizations, NIST, 2009.
          http://csrc.nist.gov/publications/nistpubs/800-53-Rev3/sp800-53-rev3-final-errata.pdf

[Sca08]   K. Scarfone et al, NIST Special Publication 800-48 Revision 1, Guide to Securing Legacy
          IEEE 802.11 Wireless Networks, NIST, 2008. http://csrc.nist.gov/publications/nistpubs/800-48-
          rev1/SP800-48r1.pdf

[WMF]   WiMAX Forum® Network Architecture Release 1.5 - Stage 3: Detailed Protocols and
          Procedure. http://wimaxforum.org/resources/documents/technical/T33

www.ingramcontent.com/pod-product-compliance
Lightning Source LLC
Chambersburg PA
CBHW081359170526

45166CB00010B/3147